THE
BIG BOOK OF
ANIMALS

DERRYDALE BOOKS

New York

Contents

The Enormous Elephant

The African elephant is the largest land animal in the world. It can be 13 feet tall to the shoulder and weigh over 6 tons. The elephant has an appetite that matches its size. It can eat as much as 500 pounds of leaves, grains, fruits and vegetables in one day.

There are two types of elephant—the Asiatic and the African. Even though the Asiatic elephant is generally smaller than the African type, the world's smallest elephant is the African pygmy. It is rarely more than 8 feet high.

All elephants have a bulky body and very thick legs. They have large ears which they move back and forth like fans, keeping them cool in hot weather. Their nose, called a trunk, is extremely long and very sensitive.

Elephants use their nose to gather water to drink or to splash over themselves, and to pull branches and leaves from trees to eat. They can warn other elephants of danger by thumping the ground with their trunk. By wrapping it around their baby they can carry the young one to safety.

A female elephant gives birth to only one baby, called a calf, at a time. The newborn is usually 3 feet high and weighs from 220 to 300 pounds. Although it can walk when it is about 1 hour old, the youngster stays with its mother for more than 2 years.

Elephants live in groups, or herds. A herd may have from 10 to 50 members, most of which are related. The herd's leader is usually an old and wise female elephant, called **a cow.**

The Cuddly Koala

Although it is not a true bear because it has a pouch, the cuddly, lovable koala bear of Australia was the model for the toy Teddy Bear. The koala is about 2 feet long and weighs as much as 30 pounds. It has round, fluffy ears but no tail.

Koalas eat only the leaves of eucalyptus trees. Oddly enough, this is their best protection against enemies. The flavor of these leaves makes the koala untasty to even the hungriest of animals.

The Slinky Spotted Leopard

The spotted leopard is the third largest member of the cat family. A full-grown male is about 5 feet long and has a 2- to 4-foot-long tail. It weighs 150 to 200 pounds. Leopards are found throughout Africa and Asia.

The leopard's coat is yellow and tan in color and is beautifully marked with flower-shaped patches of black. One rare type of leopard, sometimes called a black panther, is found only in Asia and is black all over.

Although a leopard will hunt during the day, it prefers to search for food at night. It often sits in a tree, waiting to ambush, or surprise, some animal that has come looking for water. It leaps silently from its hiding place and with its sharp teeth and claws swiftly kills its prey. The leopard may then drag its food up into the tree to enjoy its dinner alone.

10

The Delicate Deer

The graceful, swift-running deer can be found in all parts of the world, except in Australia. Although such bulky animals as the moose belong to the same family, deer are elegant animals with slender legs, long necks and large eyes.

The male deer, called a stag or a buck, has beautifully-branching bony antlers which grow from its forehead. The antlers are shed each winter and a new set begins to grow in the spring. And so, you cannot tell the age of a stag by the size of the antlers or the number of their points, as many people believe.

One or two young deer, called fawns, are born to a doe, or female deer, in the spring. The spotted coat of fawns helps make them hard to see in the grass and bushes. As they grow older the spots fade and the deer's coloring becomes the reddish or grayish brown of the adult.

The Splendid Swan

The beautiful swan is the largest member of the duck family. It has a long, arching neck and is the most graceful of the swimming birds. Because of its neck, the swan can reach far underwater to feed on weeds and seeds.

The swan commonly found in the Northern Hemisphere is white. The unusual black swan lives in Australia. A white, black-necked swan can be found in the southern part of South America.

A male and female swan stay together for life. They are very protective parents. Their babies, usually 4 to 8 in number, are hatched in a bulky nest placed near the water's edge or in shallow water.

The Giant Giraffe

The long neck and legs of the giraffe make it the tallest animal in the world. It can be 10 feet tall to the shoulder and a full 18 feet tall to the top of its head. Each of the seven bones in a giraffe's neck could be over a foot long! A newborn giraffe is a giant baby measuring 6 feet in height.

The short hair of the giraffe is colored by dark and light brown patches. Some species, or types, of giraffe have patches that fit together like the pieces of a puzzle. A giraffe's head is topped with two horns covered with velvety skin.

The giraffe's height allows it to easily gather its favorite food—the leaves of such trees as the acacia thorn and mimosa. It uses its 18-inch-long tongue to pluck them from the branches.

Extremely long front legs make drinking water or eating grass very hard for giraffes. They must widely spread their legs and bend all the way over. It must be very uncomfortable because giraffes, like camels, can go for days without water.

The giraffe's most dangerous enemy is the lion. But if it sights the lion in time it can gallop to safety at 30 miles per hour.

The Furry Fox

Foxes live in the woodlands of the Americas, Europe and Asia. These members of the dog family have a slender body with short legs, long ears and a long, bushy tail.

A vixen, or female fox, usually gives birth to her cubs in early spring. The cubs are covered with a dense coat of fuzzy fur from birth.

Many people consider foxes to be nuisances because they steal poultry and eggs. But foxes eat many kinds of small animals, including such pests as rats and insects.

Foxes can see, hear and smell extremely well and are very intelligent animals. Fox-hunting is a traditional and popular sport in some countries, especially in England. The huntsmen, on horseback, use dogs to track the fox. The winner is the person who can "outfox" the fox.

The Shifty Shark

We know that sharks first appeared in fresh water about 350 million years ago. Today there are almost 300 different kinds of shark ranging in size from 10 inches to 60 feet in length. Most now live in warm seas where the water temperature is at least 65° Fahrenheit.

Sharks are carnivorous, or meat eaters. Curiously enough, the largest sharks eat the smallest animals. Most sharks have 5 to 7 rows of teeth. As the teeth in the outer row become worn or are lost, the teeth in the second row take their place. A new row of teeth, furthest inside the mouth, begins to grow.

Although sharks are generally sluggish fish they can become very active when feeding. When sharks are so excited and are in what is called a "feeding frenzy" they have been known to eat practically anything, from rubber tires and tin cans to other sharks.

The Trotting Turkey

The turkey is a large bird. Its plumage, or feathers, is mainly blackish but it often gleams green, bronze and blue. Although the turkey is heavily feathered it has no feathers on its brightly-colored head and neck.

Turkeys spend most of their time on the ground looking for seeds and insects to eat. But if frightened they can spread their wings and easily fly to the safety of a tree.

Wild turkeys can be found only in North America. When the Spanish conquistadors arrived in the New World in the early 16th century they were the first Europeans to see a turkey, though the people of Mexico had already domesticated, or tamed, the bird.

The Topaz Tiger

Tigers live only in Asia but look different from one another depending upon the area in which they live. Tigers that live in cold places are a pale yellowish brown color with black stripes. They are also larger than tigers that live in warm areas. The Siberian tiger weighs up to 650 pounds and may be 11 to 13 feet long. Tigers that live in warmer regions are usually a bright reddish orange with black stripes. The Indian, or Bengal, tiger usually weighs from 300 to 550 pounds.

Like all cats, tigers are quick and silent hunters. Their claws and teeth are very sharp. Tigers are unusually good swimmers.

Tiger cubs, usually two or three in a litter, weigh 2 or 3 pounds at birth. When they are only 6 to 8 weeks old they go on their first hunting trip with their mother. When they are about 1 year old they begin to hunt alone.

The Silvery Seal

The gray seal has a silvery coat faintly spotted on the back and sides. It is called a true seal because it has hair instead of fur and its ears do not show on the outside.

All seals are excellent swimmers. They are mammals but have paddles or flippers instead of legs. Their body is so shaped that moving through water is easier than walking on land. Even though seals are mainly aquatic, or water, animals they rarely swim far from land. They like to climb onto a beach to bathe in the warm sun.

Seal pups, or babies, are born with a pure white, woolly coat. But after only three weeks the coat begins to darken. This is just about the time they are ready to take their first swimming lesson in the ocean.

The Chattering Chimpanzee

Chimpanzees are members of the ape family—the group of animals most like human beings. They are very intelligent and talkative animals. Scientists know that their chatter has meaning. They believe that chimps have their own special language which has at least 30 different sounds. Remember that there are only 26 letters in our own English alphabet.

Chimps live in the forests of central and western Africa. Although they often walk on all fours, chimps spend much time in trees searching for fruit. They also sleep in trees. Each chimpanzee builds a new nest each night high in a tree. The long arms and legs of chimpanzees make it easy for them to travel swiftly among trees when they are looking for a new feeding ground or a tree in which to build a fresh nest.

A chimp usually weighs about 3½ pounds at birth. Like a human, a chimp is a slow grower. It is helpless during the first year of its life and clings to its mother for guidance and protection.

Chimps grow until they are about 13 years old. At this age they are about 4 feet tall when standing upright. The average female chimp weighs about 80 pounds and the male about 100 pounds.

The Leaping Lion

The lion is a very large, strong and wild member of the cat family. An adult male can grow to have a 6-foot-long body and a 3-foot-long tail. It weighs about 500 pounds. A male lion also has a collar of long hair, called a mane, surrounding its face. The female is smaller and has no mane. The lion has powerful muscles which are used for leaping on its prey. Its sharp, hooked claws and teeth are frightening weapons.

Lions eat mainly antelope and zebra, which they usually hunt at night. They prefer to spend the hot daytime hours asleep in some shady spot.

A male, several females, and their young make up the family group called a pride. Two or three spotted cubs are born to a mother at one time. Unlike the kittens of a tame cat, lion cubs are born with their eyes open. The cubs are slow growers. They stay in the protection of the family group until they are at least 1 year old.

Although lions once lived in other areas of the world they can now be found only on the plains of Africa, in a few areas in western India and, of course, in zoos.

The Plucky Parrot

There are more than 300 different kinds of birds called parrots. All have a heavy body and a large head. Their beak is strongly hooked and powerful enough to crack the hardest nut. The short and strong legs of parrots make them the best climbers among birds.

Parrots live in the jungles of the warm areas of the world. Their favorite foods are fruits and seeds.

Some parrots are very good mimics. They learn to repeat the sounds they hear and the actions they see. Parrots are also easily tamed and trained. For these reasons parrots are popular pets.

The macaw of South America is one kind of parrot. Although most parrots are green, this bird is vividly colored in red, blue and yellow.

The Dependable Donkey

The donkey, or ass, is a close relative of the horse, although it is smaller and has ears that are 7 to 12 inches long. It has a short, bristly mane of hair growing down the back of its neck. The donkey is gray and light brown in color and has a black stripe down its back.

Donkeys are domesticated, or tame, and they are used for riding and to pull wagons and carts. The small donkeys, called burros, are used as pack animals. They can carry heavy loads on their back.

The donkey is often called stupid. In fact it is an intelligent and hard-working animal. It only becomes stubborn if mistreated.

The Observant Ostrich

The long-necked, long-legged ostrich is the largest bird in the world. An adult can be nearly 8 feet tall and weigh 300 pounds. Male ostriches are black with white wing and tail feathers. Females are grayish. Unlike most birds, the ostrich cannot fly. It uses its wings to keep balance while running. A full grown ostrich can run as fast as 30 miles per hour. And the ostrich chick is able to run almost from the moment it is born.

When facing danger the ostrich does not hide its head in the sand as many people believe. It may swiftly run to safety but it is also a fierce fighter. The ostrich will kick powerfully when cornered and use its sharp claws to badly cut its enemy.

The ostrich makes its home on the grasslands of Africa. It is a very social animal. Three to 20 birds live together in a flock. Several hen ostriches may lay their 2-pound eggs in the same nest.

The Swimming Sea Turtle

Turtles are reptiles. Like all reptiles they breathe with lungs instead of through gills like fish, and they lay eggs.

All turtles carry a hard shell on their back. This shell is made of bony plates and is attached to the turtles' backbone and ribs. Turtles can pull their head, neck and legs inside their shell. Although the shell is somewhat clumsy to be carrying about, it is very good protection against enemies.

Sea turtles have flipper-like limbs instead of feet with toes like their more land-loving relatives. Sea turtles spend most of their time in the water but come onto land to lay their eggs in a hole they have dug in the sand. The heat of the sun incubates the eggs. When the young ones hatch they somehow know to head for the protection of the sea. On land they are slow moving and can make a tasty meal for some other animal.

The Raging Rhino

Rhinoceros means "nose horn." If you have ever seen a rhinoceros then you know it is a very good name indeed. Rhinos have either one or two horns on their head, just above their nostrils. These horns are made of many tough hairs tightly pressed together.

There are five species, or types, of rhinoceros. Two types live in Africa—the white

rhino and the black rhino. The white rhinoceros, which is pale gray in color, is the largest of all rhinos. It weighs about 3½ tons and is 6½ feet high. All African rhinos have two horns on their snout. The bigger horn can measure up to 5 feet in length.

Three types of one-horned rhinoceros live in Asia. The Sumatran is the smallest rhino. It usually weighs less than 1 ton. The Indian rhino has a stubby horn only 1 foot long. The female Javan rhino has no horn at all.

A female rhinoceros has only one young, a calf, at a time. The calf remains with its mother until it is several years old.

A rhino has a good sense of smell, but cannot hear or see very well. It is not very observant of what goes on around it but if disturbed will charge, or attack, running at 30 miles per hour.

The Willful Weasel

The weasel is a slender and short animal. Its head and neck are slim and its tail is very long. This animal is fast and flexible—it can quickly disappear into a very small hiding place. The weasel is also a ferocious hunter. It kills poultry, squirrels and rodents.

The common reddish brown and white weasel lives in North America, Europe and Asia. One species, or type, of weasel which is found in cool regions turns white in winter. It is called an ermine and is trapped for its fur.

Baby weasels are born in a den hidden under a rock or boulder. They are very weak and helpless but their mother is just as ferocious when protecting them as she is when hunting.

The Powerful Polar Bear

The large, shaggy, white polar bear spends all of its life in the water and on the ice in the cold North Pole area. It has thick pads of fur on its feet, which keep it from slipping on ice.

Although a full-grown male may be up to 9 feet long and weigh as much as 1,600 pounds, polar bears have more slender bodies than other bears. Their heads are also smaller and more pointed. This shape helps them move through water without effort when looking for the fish and other animals they eat. A powerful swimmer, the polar bear is sometimes seen miles at sea, easily traveling from one iceberg to another.

Cubs, usually one or two in number, are born during the winter in a deep hole the mother digs in the snow. Because life in the cold Arctic is hard, cubs usually stay with their mother for about two years.

The Zigzagged Zebra

The zebra belongs to the horse family. It is a medium-sized animal, usually measuring about 4½ feet to the shoulder.

Unlike other types of horses, the zebra has never been tamed. It grazes in herds on the plains of Africa, eating grass and other plants. Like most horses, zebras are very fast runners. With its long and slender legs, a newborn zebra looks very much like a baby horse.

The zebra is quickly recognized because it has stripes. These are usually either black and white, or brown and cream in color. During daylight zebras can easily be seen. But at night, when their greatest enemy the lion is hunting, they are very hard to see.

The Bright Bird of Paradise

The brightly-colored bird of paradise is often called the most beautiful of all birds. With its brilliant plumage, or feathers, it is hard to believe that it is a relative of the common black crow.

Although the body length of the bird of paradise may be from 4 to 12 inches, its long tail feathers can make it look much bigger.

The bird of paradise comes in a variety of dazzling colors—black, dark blue, orange, purple, red, tan, and light blue and green. Males sometimes show off by spreading their tail feathers in a fountain-like display. Once hunted for its feathers, the bird of paradise is now protected by law.

The bird of paradise can be found only in the forests of New Guinea and its nearby islands.

The Kicking Kangaroo

The kangaroo is probably the most famous of Australian animals. It can be a strange sight to see one hopping and leaping along on its huge hind legs, and even stranger to see a baby "roo" tucked comfortably in its front pouch!

The red kangaroo is one of the largest species, or kinds, of kangaroo. It grows to be 6 to 8 feet tall and weighs about 200 pounds. Using its powerful hind legs, it can leap **as** far as 25 feet and run as fast as 25 to 30 miles per hour.

Female kangaroos usually give birth to only one baby at a time. The newborn kangaroo is blind, hairless and extremely small—only about 1 inch long. Somehow the young one knows to follow the path made by its mother through her fur and climb into her pouch. It leaves the fur-lined pouch for the first time when it is about 4 months old.

51

The Brave Boxer

The boxer is a medium-sized breed, or type, of dog. It is 21 to 24 inches high and weighs between 55 and 70 pounds. Its short-haired coat is usually a golden tan color and is sometimes flecked with gray. With their darker faces they sometimes look as if they are wearing a dark mask on their square-shaped head.

Boxers are strong, sturdy and dependable and make very good working dogs. They are also good natured and make fine household pets.

The Hefty Hippopotamus

The hippopotamus is a water-loving African animal. It has a barrel-shaped body with a huge head and short legs. Its skin is very tough and hairless. Hippos are among the largest of animals. They may be as much as 12 feet long and weigh over 3 tons.

The hippo spends most of its time in the cooling waters of swamps and rivers, often floating with only its eyes, ears, and nose above the surface. When giving birth, a female hippo sinks to the river bed, or bottom. Her baby, only 10 minutes after it is born, can swim to the surface to breathe. It isn't surprising that *hippopotamus* means "river horse."

The Building Beaver

The dark brown beaver lives in the woodlands of North America. It is a heavy-bodied animal with short, strong legs and can weigh up to 65 pounds. Like all rodents, the beaver has large front teeth.

Beavers live in fresh-water streams and ponds. They use their webbed hind feet and broad tail for swimming. Their dense, waterproof fur protects them in chilly water.

The beaver is a master builder. To build its home, called a lodge, it gathers logs and branches. With its sharp, chisel-like teeth it can cut down trees as much as 2 feet in diameter (wide). It uses its paws and paddle-like tail to tightly pack mud between the pieces of wood. To keep the entrance to the lodge underwater and safe from enemies the beaver also builds a dam across the stream. When the building is finished the dammed area of the stream or pond becomes the home of several beaver families.

57

The Handsome Horse

The horse has a sleek, well-rounded body with long and slender legs. Its neck is also long and arches gracefully. The horse's head tapers into a rounded muzzle and ends in a velvety nose. A mane of long hair grows down the back of its neck and a lock of the same color lies between its pointed ears.

The different types of horse vary greatly in size, from the miniature pony to the giant Clydesdale. Horses also come in many shades and patterns of black, brown, tan, cream, white, gray, red and gold.

Horses were probably first domesticated, or tamed, about 5,000 years ago. Since then people have trained them to do many different kinds of jobs—pulling wagons, racing for sport, rounding up cattle and horseback riding for pleasure. They are also good companions.

A female horse, called a mare or a dam, usually gives birth to only one young, or foal, at a time. As it grows, the male foal is called a colt and the female a filly. The male horse is called a stallion.

The Delightful Dolphin

The bottlenose dolphin lives in the sea and looks like a fish but is a mammal because it has lungs and does not lay eggs. It has thin, powerful tail fins and can swim as fast as 20 miles per hour. The dolphin is 8 to 9 feet long when full grown.

Dolphins can be tamed and are friendly, playful animals. Schools, or groups, of dolphins sometimes travel alongside ships in the ocean or guide them into harbor.

Dolphins are very intelligent. We know that they can talk to one another using a language of special sounds. Scientists are studying their language but we do not yet understand it. Perhaps, someday, you will be able to talk with a dolphin!

The Brawny Brown Bear

All bears have a thick covering of hair and very short tails. Most have heavy, bulky bodies and thick, short legs. They range in length from 3 to 8 feet and weigh between 50 and 1,500 pounds.

The European brown bear is usually found living alone in the deepest part of the forest. Even though it is a very strong bear and fairly large—it weighs about 800 pounds—it is usually a gentle animal.

During the summer the brown bear eats many kinds of vegetable foods, fish and small animals. A thick layer of fat builds up under its skin. This fat provides the energy the bear needs to stay alive during its winter hibernation in a cozy cave.

Either one or two, but usually twin, brown bear cubs are born in December or January, while the mother is hibernating.

The Downy Duck

There are almost 150 species, or types, of duck. They live in all parts of the world but most can be found in the cooler northern regions. They range in size from 1 to 2 feet in length.

Ducks are aquatic, or water, birds. They have a compact body, short legs and webbed feet. Ducks are excellent swimmers and often dive underwater for food. Their feathers, or plumage, are very dense and shed water easily. A layer of fine, soft feathers, called down, lies close to their skin and keeps ducks especially warm. Ducks are also strong fliers. During winter many species migrate long distances to warmer areas.

The white ducks we see most often on farms are descended from the mallard, which is an easily tamed wild duck.

The Lanky Lynx

The lynx is a short-tailed wild member of the cat family. It is found in the cooler regions of Europe, Asia, Africa and North America. The true lynx of North America is called a Canada lynx.

The Canada lynx lives in the forests of Canada and the northern United States. Its coat is faintly spotted and its tail is tipped with black. The broad, furry feet of the lynx act like snowshoes, helping the lynx to move swiftly over snow in search of rabbits and other small animals. Like most cats, the lynx prefers to hunt at night.

The Prickly Porcupine

The porcupine is a large rodent with short legs. Old World porcupines, found in India, Africa and southern Europe, are about 2½ feet long. They have thick, soft fur and many long, sharp, needle-like quills for protection against their enemies. Their quills are loosely attached and will fall out when touched.

The New World porcupines are found in both North and South America. They may reach 3½ feet in length and weigh as much as 40 pounds. They also have spines but their hair is long and coarse. A single animal may have as many as 30,000 quills! The porcupine lives in wooded areas where it is sure to find plenty of its favorite food—the leaves and bark of trees.

The Fiery Flamingo

The flamingo is a tropical bird. It can be found mainly around the Mediterranean Sea and in Asia, South Africa and the American tropics. It was once plentiful in Florida.

The feathers, or plumage, of the flamingo come in lovely shades of pink, white and rose. Some of the wing feathers are black.

The flamingo is one of the tallest birds. With its long legs and neck it can be as much as 4 feet tall. When searching for food the flamingo wades, or walks through shallow water, with its head below the surface. In this position the flamingo's downward-curving bill is upside-down. It acts as an efficient scoop, collecting shellfish from the botton.

The Bristly Wild Boar

Boars are members of the swine family. This wild pig can be found throughout Europe, Asia and North America.

Wild boars are large animals. They can measure up to 6 feet in length including their 12-inch tail, and stand 40 inches in height to the shoulder. Males can weigh as much as 450 pounds while the smaller female weighs about 300. The blackish brown coat of boars is made of stiff bristles and softer, curly hair. An adult's tusks can grow to be 12 inches long and are deadly weapons.

Wild boars root, or dig, in the ground with their sensitive snout looking for roots and bulbs to eat. They also eat nuts and fruit that have fallen to the ground.

The rough stripes of wild boar piglets make it hard to see them in dense under- brush. These stripes fade away when the piglets are about 6 months old.

The Soaring Seagull

The long-winged seagull is found living near seacoasts throughout the world. Though it sometimes flies inland to marshes or lakes it is rarely seen far out at sea.

The gull most commonly found in North American and Eurasia is gray and white. It is 22 inches long when full grown, weighs about 2 pounds, and measures about 4 feet from wing tip to wing tip. With its great wingspan, the gull is one of the most graceful fliers.

The bill of the gull, which curves downward at the tip into a sharp hook, helps it to feed on fish and shellfish. Gulls are sometimes considered a noisy nuisance, but they are also good scavengers, finding food in the garbage dumped in harbors and coastal waters.

The Tracking Timber Wolf

The timber, or gray, wolf of Europe and North America is a wild member of the dog family. Like the dogs people keep as pets, they are intelligent and can smell, see and hear very well. They are also excellent hunters.

Wolves hunt in family groups called packs. They may travel throughout an area of several hundred square miles searching for food. They look for small animals such as rabbits and fish. But by working together they can capture such large animals as moose and deer. If they find no prey they will eat the meat of dead animals, or even fruit.

The loyalty and affection among family members is very strong. A male and female wolf will often stay together for life and raise many litters of cubs.

The adult male gray wolf weighs about 150 pounds. From the tip of its nose to the end of its long and bushy tail it measures about 6 feet long.

The Perky Penguin

Have you ever heard of a bird that can fly through water but not through air? The penguin can! It is streamlined in shape and its tightly-packed feathers help it to shed water and keep warm in freezing ocean water. Its wings are small and paddle-like. With its flipper-wings it can travel as fast as 25 miles per hour.

Penguins spend most of their life in the water, coming onto land only to lay eggs. Because penguins feed only underwater,

they spend months without eating, waiting for their young to hatch and grow. On land, penguins walk upright and waddle slowly along. On slippery snow and ice, they flop on their belly and smoothly "row" with their feet.

Penguins are found only in the Southern Hemisphere of the world, from the cold Antartic region, north to the coasts of Australia, New Zealand, South America and southern Africa.

The Fleecy Sheep

Most sheep are short, stocky animals with thick legs and a dense, woolly coat called fleece. They weigh between 100 and 350 pounds. Flat surfaces on their side teeth help them to chew the grass and other plants they eat.

Sheep have been domesticated, or tamed, for thousands of years. At first they were probably kept for their meat, milk and skins. Though we do not know when, the idea of shearing, or cutting, a sheep's wool and turning it into cloth came much later. As much as 20 pounds of wool can be cut at one time from a single sheep.

The horned male sheep is called a ram, the female a ewe and the young sheep a lamb.

The Playful Pony

A pony is a type of horse that is rather small when full grown. It is usually from 32 to 52 inches in height when measured from the ground to the animal's shoulder. Although small, ponies are very sturdy animals.

The Shetland is the most famous breed, or type, of pony. It is both strong and gentle. In some places these ponies are used to pull wagons and do other work. The Shetland also makes a good pet and is easy for children to ride. Have you ever ridden a pony?

The Optimistic Otter

Otters are graceful, rapid swimmers and spend most of their time in or near water. Their dense, shiny fur keeps them warm in cold water and their skin dry. They expertly use their webbed feet, strong legs and thick, rudder-like tail to move quickly through water, even though they are somewhat clumsy on land.

The otter is found living in every part of the world, except Australia. The North American otter is about 3½ feet long and weighs about 20 pounds. It makes its home in the bank of a river or in a nearby hollow log and searches the water for fish, crayfish and frogs.

A playful and curious animal, the otter seems to enjoy sliding down river and snow banks and swimming—just for the fun of it.

The Pearly Pelican

The pelican is a large water bird that is usually white or pale gray in color. Its body can be as much as 5 feet long. Some kinds of pelicans live near inland fresh-water lakes. Others prefer to live near salt water and can be found along seacoasts and on islands.

Hanging from the flat beak of the pelican is a large pouch of featherless skin. When feeding, the pelican uses this pouch to scoop fish from the water. Later it swallows its catch.

Pelican parents take turns sitting on their nest of eggs and both care for the young when hatched. Baby pelicans are born without feathers, but soon grow a soft, fluffy covering of fine feathers called down.

The Crafty Crocodile

Most of the 25 species, or types, of crocodile live in the warm parts of North and South America, Africa, Australia and Asia. Some types live near salt water but most prefer fresh-water rivers, lakes and swamps. Crocodiles are excellent swimmers, using their powerful tail to move swiftly through water.

The triangle-shaped head of the crocodile has a long tapering snout. The large, strong jaws hold many sharp teeth. A crocodile can kill animals as large as cattle and deer.

Although they never stop growing, crocodiles usually average 12 feet in length. A crocodile can weigh as much as 500 pounds and live for 20 to 30 years.

89

The Tropical Toucan

The toucan of South America is a noisy jungle dweller with a large canoe-shaped and brightly colored bill. Toucans sometimes use this bill to fence playfully with one another, but the bill is not used as a weapon.

Toucans do not have very strong wings and can fly only for short distances. Traveling in small flocks, or groups, they leap among treetops searching for insects and fruit to eat.

The Beautiful Black Leopard

The rare black leopard, or black panther, lives in Asia. It is black all over and slightly bigger than its relative the African spotted leopard.

Leopards have short, powerful legs. With their long and sharp claws they can easily climb trees. When hunting they will often sit on a large tree limb and wait to leap upon deer or other animals. The leopard usually hunts alone and at night.

Panther cubs are born in the safety of the deepest part of the jungle. When they are a few months old they begin learning how to get their own food and survive in the wild jungle.